HOME JUNGLE

Living with plants

27 STYLISH
AND NATURAL
INTERIOR
DESIGNS

SONIA LUCANO

Photographs
FRÉDÉRIC LUCANO

GINGKO PRESS

Sonia Lucano
Designer and artist

After studying at the Musée des Arts décoratifs in Paris, Sonia Lucano began her career as a fashion designer. Her passion for interior decor, however, means that she now divides her time between these two contrasting worlds, alternately arranging photo shoots for books and magazines and creating *objets d'art*.

An inveterate collector, Sonia is always on the lookout for new materials to combine, assemble, decorate, and transform.

She has written several books on interior design and DIY.

Sonia lives in Paris with her husband, photographer Frédéric Lucano, and their three children.

CONTENTS

LIVING WITH PLANTS

The idea for this book came from my overwhelming desire to escape to the country after 20 years of living in Paris…but without actually moving! I wanted to stay in an urban environment while transforming my home into a green space.

There are three key starting points: Think green; be eclectic in your choice of plants; and make use of whatever materials you have on hand.

The main thing, I think, is not to restrict yourself when selecting plants. Try anything, even—I should say, especially—if you thought it had gone out of fashion: plants you always see on desks or in waiting rooms, or plants your grandparents used to have. Plants you hardly notice any more. And be daring: Use cacti and succulents by all means, but combine them with herbs, ferns and ivy, dried leaves, even weeds…

And when it comes to bringing them to life in your home, you'll find lots of ideas and easy-to-follow instructions here in this book. You can attach them to the wall, hang them from the ceiling, put them in unusual containers, and apply all kinds of decorative techniques—painting, knitting, macramé, sewing, origami…Anything goes!

LEGEND OF PICTOGRAMS

CREATION	CARE	COST
❩❩❩❩❩	❩❩❩❩❩	❩❩❩
from easy to tricky	from easy to tricky	from low to moderate ($15–50 / 10–50€)

01
02
04
05
06
07
08

03

What you need

You don't need much to get started on your green-living adventure. I've listed the basic items below. Any specific requirements are included in the instructions for each design.

01 – Scissors

02 – Indoor plant compost

03 – Newspaper

04 – A small watering can

05 – Secateurs

06 – Gloves
(for handling cacti)

07 – String

08 – A spray bottle

PLANTS

CACTI AND SUCCULENTS

(01–06)

Succulents are fleshy plants that store water, enabling them to survive in extremely dry conditions. The surface of the plant is waxy to prevent the water from evaporating. Cacti are just succulents with spines or hair.

Succulents normally live in areas where there's no rain for long periods, so try to replicate that type of environment in your home. In the autumn, water succulents (other than cacti) no more than once a month, and don't water them at all in the winter, but make sure you put them somewhere where there's plenty of light. When spring arrives, and the outdoor temperature rises, water them every two weeks. This will 'wake them up' and, once they're fully stocked with water again, they'll grow and flower.

More or less the same applies to cacti, which are dormant from mid-October to mid-March and don't need any watering, unless the air in your home is very dry—in which case, give them some water once or twice during the winter. Between March and October, water them once a month. Be generous when you do, but make sure no water collects in the saucer, as cacti don't like having wet feet. Avoid this by emptying the saucer 15 minutes after watering.

RETRO PLANTS

(07–12)

By 'retro plants' I mean anything that your grandparents might have had in their homes. You might not even think of using them because they'll look old-fashioned. Well, think again! Old plants are back in style—and what's more, they're tough and don't need much looking after.

Here are four 'essential' retro plants:

Areca palm ***(Chrysalidocarpus lutescens)*** *(07)*, also known as the golden cane, yellow, or butterfly palm.

Pothos plant ***(Epipremnum aureum)*** *(08)*, also known as golden pothos or devil's ivy. With their delicately drooping stems, pothos plants will give your home a bohemian feel. They don't like too much water and can rot, so water them once a week in winter and only a little more in summer. The rule of thumb is to let the compost dry out completely before watering them again.

Swiss cheese plant ***(Monstera deliciosa)*** *(09, 10, 11)*, also known as the ceriman or split-leaf philodendron (although strictly the latter is a slightly different plant). Easily recognizable by their mass of large, shiny, 'cut-out' leaves, swiss cheese plants are easy to grow—and grow quickly. They're of tropical origin, so they need watering regularly, though moderately—and less in winter. In their case, avoid letting the compost dry out, and make sure they get plenty of light.

Yucca *(12)* This is a perennial shrub that doesn't like dry heat or cold. Keep it away from radiators and water it moderately, once a week.

All these plants are green in more ways than one: They'll improve the air quality in your home, and they're especially good at absorbing carbon monoxide.

01
02
03
04
05
06
07
08
09
10
11
12

'FREE' PLANTS

(01–03)

By 'free' plants, I mean plants you can grow yourself, from seeds or cuttings. Not only will they cost you virtually nothing, they will also take you back to your childhood.

Try planting an avocado seed, for example, to see if a shoot appears, and then those pretty leaves.

If you have some potatoes that have been kicking around and are starting to sprout, the shoots should develop into leafy stems (unless they've been treated with sprout inhibitors!). Sweet potato plants have particularly attractive, heart-shaped leaves.

If you remove the crown from a pineapple and put it in water until roots appear, you can grow your very own pineapple plant.

You can also cut ivy branches or pull up brambles when you're out walking and put them in water. If they develop roots, they can then be planted.

Of course, not all of your attempts at growing seeds or cuttings will be successful, but that's okay—just don't give up!

EPIPHYTES

(04)

An epiphyte isn't really a plant at all; it's an organism with no roots that grows without needing to be in contact with soil or compost. The most common epiphytes are bromeliads (*Bromeliaceae*), which can be found in most garden centers. Some, like *Tillandsias*, literally live on fresh air and can simply be suspended in space (which is why they're known as 'air plants'). Others are more like orchids and need to be embedded in moss or bark.

Epiphytes come in all shapes and sizes. Some are almost geometrical while others have 'tentacles' (my favorites); some are hairy, some spindly, and some look like the crown of a pineapple.

Most epiphytes have pretty, silvery leaves, and they're all easy to look after: all you have to do is spray them with water once a week or so—but make sure you spray the underside of the leaves as well as the top. They don't like too much water, and if they aren't getting enough, they'll soon let you know: The tips of the leaves will turn dry and brown. If this happens, just give them another spray.

DRIED PLANTS

(05, 06)

Dried plants and foliage can be highly decorative. For example, you can simply put them in a vase to create a permanent display, frame them like a picture, or weave them into a wreath or crown.

You can pick up fallen leaves or branches in the woods (maple leaves and silver birch twigs make a particularly attractive display), or you can buy dried plants from a florist: eucalyptus is a safe bet, but also try dried *Gypsophilia* (baby's breath) and hydrangea flowers...

HERBS

(08, 09)

Herbs not only smell and taste good, they almost all look good, too, and they should feature prominently in your home—and not just on the kitchen windowsill. Most herbs are annuals, so you'll have to buy new plants each year, but some—such as mint, rosemary and thyme—are perennials. Others that you can easily grow in a pot or trough are aniseed, basil, chervil, chives, coriander, parsley, and tarragon.

Herbs need plenty of light, but don't put them near a window that gets direct sunlight, or they'll burn. They also need to be kept moist, so give them a little water every other day to prevent the soil from drying out.

01
02
03
04
05
06
07
08
09

Plants ON WALLS

a
made with love

WREATH OF DRIED LEAVES

Leaves are just as beautiful when they're dry, and this eucalyptus wreath will create a delightfully rustic effect.

1 Soak the eucalyptus branches in water while you make the wreath so that they're nice and flexible and can be bent into shape without snapping.

2 Make a wire ring about 1' (30 cm) in diameter, twisting the ends together so that it keeps its shape.

3 Cut the bottom off the eucalyptus branches and use only the top 2" (5 cm). Do the same with the thistle stems and the *Gypsophilia*.

4 Work in one direction only, to give your wreath a sense of flow. Tie the cotton thread around the base of a eucalyptus branch and wrap it around the wire ring several times along its length. Do the same with the other branches, alternating them with the flower sprays and thistle stems and attaching each one about 1½" (4 cm) further around the ring. When you get back to where you started, overlap the first branch with the last, cut the thread, and tie it off securely.

5 Make a label (see p. 90) with letter stamps and wire it to a protruding branch.

6 Tie a length of nylon thread securely to the wreath and hang it up. In a few days, the leaves will have dried and your wreath will be ready to display.

CREATION:

CARE:

COST:

MATERIALS

- About 6' (2 m) of medium-gauge wire
- Scissors, Wire cutters
- Strong green cotton thread
- A white tie-on label
- Letter stamps
- A black ink pad
- Nylon thread

PLANTS

- 8 freshly cut eucalyptus branches
- sprays of small white flowers (e.g. *Gypsophilia*)
- 4 thistle stems

FRONDS IN A FRAME

CREATION: ❩◌◌◌◌
CARE: ❩❩❩◌◌
COST: ❩◌◌

Remember pressing flowers between the pages of a book? Here you're going to flatten your favorite leaves and exhibit them—complete with botanical information.

MATERIALS

- 4 sheets of picture glass 8 × 12" (20 × 30 cm)
- 2 sheets of picture glass 12 × 16" (30 × 40 cm)
- A roll of adhesive aluminum tape
- A ruler and a cutter
- A permanent marker
- Plant labels

PLANTS

- A selection of attractive leaves (ferns, ivy, asparagus, etc.)

1 Choose the leaves you want to frame and lay them flat between the pages of a heavy book. Pile some more books on top of that one and leave them for 2–3 weeks, by which time they should be dry—and, having been kept in the dark, have retained most of their color.

2 Clean the sheets of glass with an appropriate product, taking care not to cut yourself. Then lay one flat, place the selected leaf or leaves in the center (or off-center), with a label if you want to include one, and cover with another sheet of glass the same size. Make sure the two sheets of glass are exactly aligned.

3 Cut strips of aluminum tape 1" (2.5 cm) longer than the sides of the glass sheets (i.e. 9" (22 cm) and 13" (32 cm) long for the 8 × 12 (20 × 30 cm) sheets). Then, using the ruler and cutter, cut these lengthwise into strips ⅔" (1.5 cm) wide. (Aluminum tape is usually 2" (5 cm) wide, so you will end up with three strips per length.)

4 With the marker, make marks one-third of the width of your strips of tape from the edges of the lower sheet of glass.

5 Remove the backing from the strips of aluminum tape and, looking down through the top sheet of glass, line them up with the marks underneath. Press them down gently with your finger (they stick well to glass), fold the remaining tape over the edges of the glass, and stick it to the lower sheet. Please use extreme care in handling both the glass and aluminum tape to avoid cutting your finger.

HERBIER DE

CLIMBING WALLPAPER

CREATION: ●○○○○
CARE: ●○○○○
COST: ●○○

Pothos plant stems can be several yards long and make pretty patterns on a wall or an attractive frame for a doorway, adding a touch of wildness to your interiors.

MATERIALS

- About 30 small nails
- A hammer
- A roll of thick nylon thread

PLANTS

- A potted pothos plant

01 Choose the wall you want your ivy to climb up—ideally one with a feature, like a doorway or a table or desk.

02 Hammer small nails into the wall, about 15–20" (40–50 cm) apart, along the lines you want your ivy to trace, and join them with the nylon thread, winding this once around each nail.

03 Place your plant on the table or desk, or simply on the floor, and 'show it the way' by twining the branches around the threads.

Pothos plants are natural climbers and grow very quickly. They need regular watering (about twice a week) and plenty of light, but not direct sunshine. They can also do well in a shady spot, yet the color of their leaves will vary according to the amount of light they're exposed to: yellower if well lit, dark green if kept in the shade.

PLANT-PRINT FABRIC

CREATION: ❩❩❩❩❩
CARE: ❩❩❩❩❩
COST: ❩❩❩

MATERIALS

- 2 rectangular pieces of unbleached cotton fabric (ca. 14 × 20" / 35 × 50 cm)
- An iron
- Khaki or olive-green fabric paint and a small paintbrush
- Paper towels
- A rolling pin
- A sewing machine
- Cushion stuffing
- A needle
- Unbleached cotton thread

PLANTS

- 5 or 6 freshly cut leaves

Create an impression of plant life in your home by printing leaf designs onto fabric and either framing it or making it into a cushion.

1 To print the fabric:

- Wash the fabric to remove any coating, and iron it.
- Lay your selected leaves in the pattern you want to create on the 'right' side of one of the rectangles of fabric, allowing some of them to overlap the edge.
- Taking one leaf at a time, coat the upper surface thinly with paint and lay it back down onto the fabric. Cover it with a paper towel, and press it down firmly with the rolling pin, making sure that the leaf is pushed right into the fabric. When you remove the paper, the leaf should come away with it.
- Repeat the procedure with each leaf, being careful not to overlap a previous print if the paint is still wet. Leave the fabric to dry completely. Then iron it (without steam) to seal the paint.

2 To make the cushion:

- Oversew all four sides of each rectangle of fabric.
- Lay the two rectangles one on top of the other, 'right' side to 'right' side.
- Sew all the way around the outside of the rectangles, about 0.4" (1 cm) from the edge, leaving a 6" (15 cm) opening on one side.
- Turn the fabric right side out and stuff it.
- Fold over the edges of the opening by 0.4" (1 cm) and sew it closed by hand.

Le Myosotis

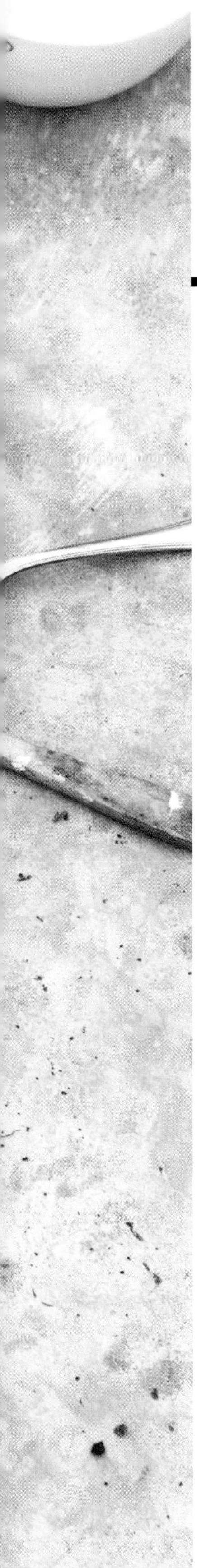

GREEN GRAFFITI

Decorate your walls with messages in moss—a natural alternative to spray paint.

CREATION: ❩❩❩
CARE: ❩❩❩
COST: ❩

MATERIALS

- A printer and paper
- Scissors
- A mixing bowl
- 1 cup of flour
- 1 cup of beer
- 1 pot of natural yogurt
- 2 soup spoons of sugar
- A saucepan
- Chalk
- Strong spray glue
- A small paintbrush
- A spray bottle

PLANTS

- A tray of moss (see p. 26)

01 Print your message on sheets of paper and cut out the letters.

02 To make the glue: Pour the flour into the bowl and gradually add the beer, stirring continuously to prevent lumps from forming. Then add the yogurt and sugar and mix well. Pour the mixture into the saucepan and put it over a low heat for 10 minutes, then let it cool. Your glue is ready to use!

03 Choose an exterior wall for your message—preferably one that gets little sun and that has a porous surface the moss can cling to (a brick wall is ideal). Hold your letters against the wall and outline them with chalk.

04 Stick the paper letters to the moss with the glue spray (not your own glue mixture!) and cut out the moss letters with scissors. Don't worry if you have to divide each letter into smaller pieces; the joins will be invisible once you've finished. Then pull off the paper letters.

05 Using your brush, apply your home-made glue liberally to both the moss letters and the wall, then affix them, pressing down firmly, making sure they stick properly. When you've stuck all your letters to the wall, give them a good spray with water.

06 For the first few days, spray your message regularly to keep the moss moist while it roots itself into the wall. After that, spray it from time to time, especially in dry weather.

To keep your message nice and clear, trim the moss every now and then with a pair of scissors.

urban

ungle

IVY TRELLIS

CREATION: ●○○○○
CARE: ●○○○○
COST: ●○○

Create a living tableau on an interior wall with this ivy trellis, and watch it weave an ever-changing picture.

MATERIALS

- A roll of medium-gauge wire
- A roll of adhesive aluminum tape
- 20 or so small nails
- A hammer

PLANTS

- Ivy in a pot

01 Cut about 25 lengths of wire 12" (30 cm) long and crimp them with the pliers every 2" (5 cm), leaving 0.8" (2 cm) 'spare' at one end. Then bend each wire into a hexagon, overlapping the last 0.8" (2 cm) (see fig. 1 on p. 86).

02 'Seal' the first hexagon with aluminum tape.

03 Then link the second hexagon to the first before sealing it, and so on, creating a linked trellis of hexagons.

04 Decide where to put your trellis (preferably on a shady wall—see below), hold it against the wall and hammer in a nail right at the top to hang it from. Then hammer in nails all along the top and here and there down the sides to keep the trellis in place.

05 Place the pot of ivy beneath the trellis and wrap its stems around the lower hexagons. As it grows, you may need to guide it around the other hexagons.

Ivy grows fast and needs plenty of water, so make sure you water the pot two or three times a week. Ivy doesn't need much light, however, and will even grow well in quite a dark room.
To create a really leafy tableau, don't be afraid to chop off any spindly stems and prune others to make them divide.

TUBEREUSE

Hanging PLANTS

GREEN DREAM-CATCHER

Suspend a string of hearts from the branch of your favorite tree.

CREATION: ●●○○○
CARE: ●●●○○
COST: ●○○

MATERIALS

- A large ball of natural jute twine
- Scissors
- A branch about 20" (50 cm) long
- A ruler
- A small plant pot (3–4" / 8–9 cm across), ideally with a matching saucer

PLANTS

- A *Ceropegia* woodii 'string of hearts'

01 Cut 30 lengths of string, each 10' (3 m) long.

02 Attach each length to the branch using a cow hitch (or lark's head knot), as shown in fig. 2A on p. 87. This will give you 30 pairs of 5' (1.5 m) strings hanging from the branch. Spread the pairs of strings along the branch, approximately 0.6" (1.5 cm) apart.

03 Following the instructions beneath fig. 2B on p. 87, create your macramé dream-catcher.

04 Place the Ceropagia in the center and pass the stems carefully through the strings. Don't break them—they're fragile!

Ceropagia *is a succulent, so it doesn't need much looking after—just a little water every 10 days or so. (Hold a bowl under the pot when you water it if you haven't managed to fix a saucer under it.) Try to keep the compost just moist all the time and never let it dry out. Ceropagia doesn't mind plenty of light and even direct sunshine, but don't put it right up against a window.*

QUIRKY KOKEDAMA

In Japanese, koke *means 'moss' and* dama *means 'ball.' Put them together and you get a clever way of growing plants without a pot.*

CREATION:))))))
CARE:)))))
COST:)))

MATERIALS

- 2 handfuls of akadama (bonsai soil)
- 2 handfuls of indoor plant compost
- Thick black or green cotton thread
- Nylon thread

PLANTS

- A few small climbing or trailing plants, e.g. ferns, asparagus, or ivy
- 5 or 6 sheets of moss

01 Pick a suitable plant: Start with a small one, which will be easier to work with. Take it out of its pot and remove as much soil as you can from the roots. If there are a lot of roots, trim them; otherwise, you will end up with a *kokedama* that's bigger than your plant!

02 Prepare the moss by removing any small pieces of wood, cutting out any brown patches, and trimming it if it's too thick—but be careful not to cut it too much, or it will literally fall to pieces.

03 Now make the *dama*: Mix the dry akadama granules with the same volume of water, leave them to absorb the water, and stir. Add granules or water as necessary until you have a mixture that holds together and can be molded into balls. Then add the same amount of plant compost and work the two substances together into a homogeneous '*kokedama* compost'.

04 Hold your plant in one hand and, with the other, apply lumps of *kokedama* compost to the roots until they're completely covered, then mold the compost into a smooth ball.

05 Stick the moss to the ball, using as many sheets as you need to cover it completely. Then wrap the string four or five times around the *kokedama* to keep the moss in place, tying it off with a double knot. Conceal the string in the moss and trim off any moss that's sticking out.

06 As soon as your *kokedama* is finished, water it generously, because the roots of your plant have been exposed to the air and started to dry out.

07 Finally, hang up your *kokedama* using the nylon thread or simply place it on a (preferably painted) surface—as shown on the following pages.

Kokedamas *need to be kept moist. Every couple of weeks, take them down and hold them directly under a faucet. In between those times, spray both the moss and the leaves of the plants lightly with water. Test the balls by squeezing them gently: They should feel just damp. Never let them dry out or your plants will turn yellow.*

NATURAL MACRAMÉ

You don't need to be a sailor to master the few knots needed for this macramé hanging basket with its cascade of green leaves.

CREATION:

CARE:

COST:

MATERIALS

- A ball of medium-gauge natural twine (jute or hemp)
- Scissors
- A ruler
- A small plant pot (4–5" / 10–13 cm across), ideally with a matching saucer

PLANTS

- A small succulent such as a *Ceropegia* or *Peperomia*

01 Make the macramé slings using the instructions on pages 88–89.

02 Hang the potted plants inside them and thread the stems carefully through the strings, taking care not to break them.

Like Ceropegia, Peperomia *is a succulent and needs little 'aftercare.' Both plants like plenty of light and even direct sunshine, but don't hang them right up against a window. Give them a little water every 10 days or so to keep the compost slightly moist, but never wet. If your pot doesn't have a saucer, hold a bowl under it when you water to catch the drips.*

POTTED BIRD CAGE

CREATION:)))))
CARE:)))))
COST:)))

No need for a bird—bring that lovely cage alive by turning it into a miniature hanging garden.

MATERIALS

- 3 small earthenware plant pots (3–4" (8-10 cm) across)
- A small bag of indoor plant compost
- A decorative bird cage (from a home decor or antique store)

PLANTS

- 3 small plants, e.g. succulents, ferns, or ivy
- Some moss

)1 Choose a variety of plants and transplant them into the earthenware pots, which might be reclaimed, found, or bought new or second-hand.

)2 Carpet the cage with moss to absorb any water that runs out of the pots.

)3 Place the pots inside the cage and gently pass the stems of the plants through the bars. As the plants grow, they'll wind their way around the bars.

)4 Finally, hang the cage from a beam or a bracket.

Water your hanging garden regularly—once or twice a week, depending how warm the room is. The compost should always be moist. Keep the moss moist, too, by spraying it just as often.

PLANTS WITH A SWING

Lend your interior that Swinging Sixties flair by letting your favorite plants 'hang out' together.

CREATION:)))))
CARE:)))))
COST:)))

MATERIALS

- A flat piece of old wood (about 20 × 8"/ 50 × 19 cm and 1"/3 cm thick)
- 1 or 2 sheets of medium sandpaper
- A drill
- A large ball of undyed string
- Scissors
- A ruler
- Glue
- 2 or 3 assorted earthenware plant pots

PLANTS

- Various small plants, e.g. *Fittonia* (nerve plant), ivy, or climbers

01 Sand down the edges of the piece of wood to prevent splinters, but leave it 'rough'-looking. Then drill four fairly large holes about 1" (3 cm) from the corners.

02 Cut 8 pieces of string, each about 26' (8 m) long.
Use four pieces for each side, threading them down through one hole in the wood, passing them underneath, and then threading them up through the other hole at the same end.

03 On each side, tie left- or right-facing square knots (see p. 88) in each of the four strings, using two central strings instead of one, for a length of 6" (15 cm). Then bring all eight strings together on each side and continue the square knots for another 6" (15 cm), using four central strings and two on either side.
Now, leave a gap of 4" (10 cm) before a further 6" (15 cm) of knots, then a gap of 2" (5 cm), then another 6" of knots, another 2" (5 cm) gap, and finally 3" (8 cm) of knots.
With the remaining length of strings, make a loop on each side. Cut two separate pieces of string about 16" (40 cm) long and wrap them about 10 times around each 'bundle' of eight strings, knotting them each time, to fasten them tightly together. Add a little glue to each end of the wrapped string to make sure it doesn't come loose.

04 Hang up your swing and arrange the potted plants on it.

These kinds of plants need to be in moist soil, so water them in small amounts and often.

LIVING MOBILE

CREATION: ❩❩❩❩❩
CARE: ❩❩❩❩❩
COST: ❩❩❩

Let your 'air plants' fly and make your interior light and graceful.

MATERIALS

- A roll of fine nylon cord
- A large needle
- 10 gold crimp beads
- A pair of fine pliers
- 4 or 5 glass globes such as those shown on pp. 46–49 (available from craft stores and places selling Christmas decorations)

PLANTS

- 4 or 5 epiphytes

❩1 Cut pieces of nylon cord long enough to hang your plants at the heights you want them. Thread them through the needle and 'sew' through each plant from side to side near the base so that there's about 4" (10 cm) of cord protruding on one side and the rest on the other.

❩2 Thread both ends of the cord through two crimp beads. Slide one of them to within an inch of the plant and squeeze it with the pliers to clamp the cord. Make a 2" (5 cm) loop at the end of the longer cord, slide the other crimp bead over the end and clamp it in the same way.

❩3 Pass each cord through the small hole in a glass globe, pulling the plant up into the globe and hanging the loop over a hook or nail.

Spray each plant with water about once a week.

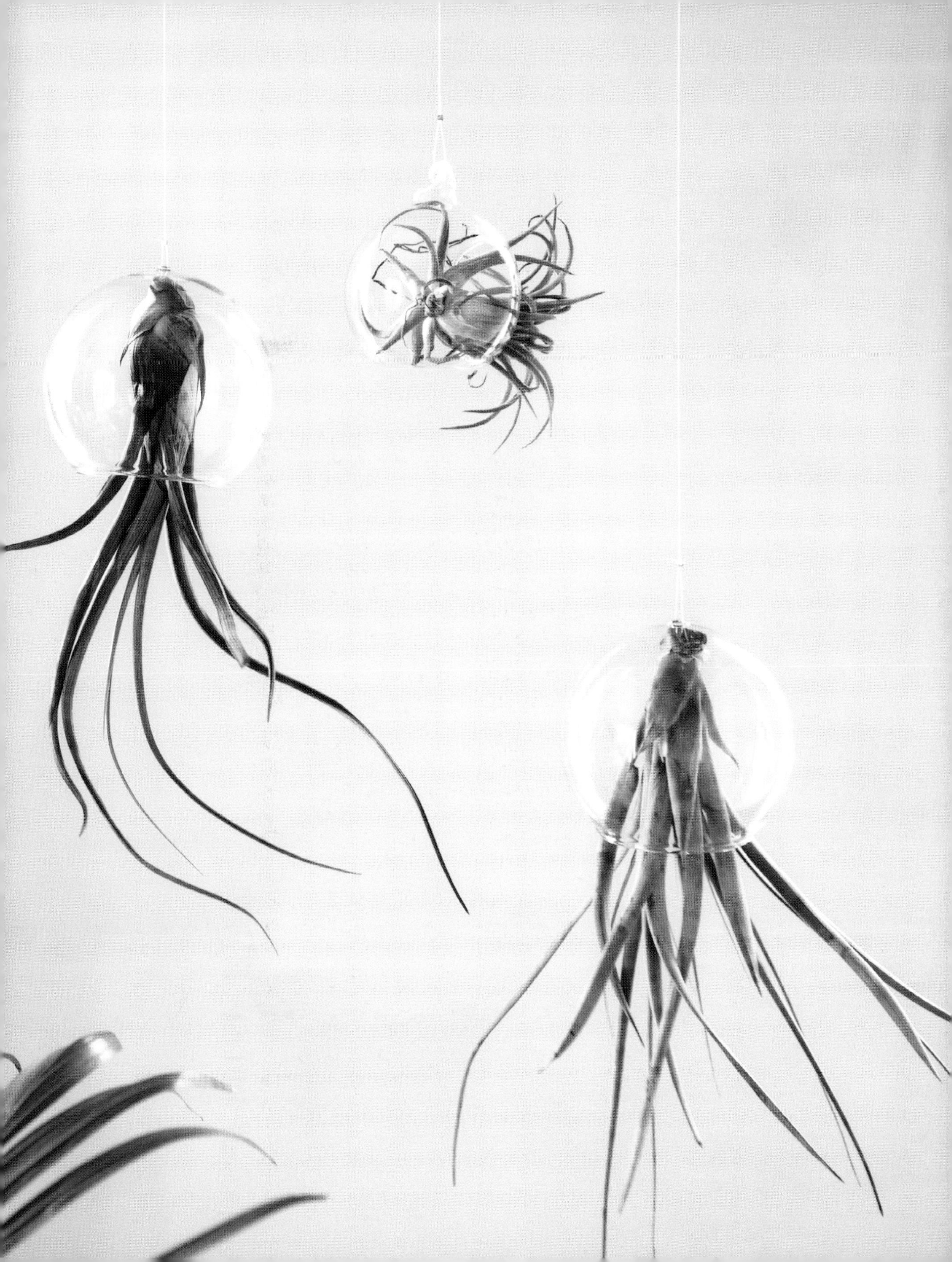

Plants IN POTS

POT COZY

Brush up on your backstitch to create funky knit vests for your plant pots.

CREATION: ❩❩❩❩❩
CARE: ❩❩❩❩❩
COST: ❩❩❩

MATERIALS

- A ball of medium-gauge undyed string
- Two 10 mm (size 15) knitting needles
- 1 very large sewing needle
- 1 terracotta pot (8" (20 cm) across and (8" (20 cm) high)
- A bag of indoor plant compost

PLANTS

- 'Retro' plants, e.g. Swiss cheese plant (*Monstera deliciosa*), Schefflera, fig, or yucca

01 Pull out about 8" (20 cm) of string (or wool) and knit a row of 40 loose stitches. Then backstitch 59 more rows, which should give you a piece of knitting around 5" (13 cm) deep.

02 With each extra row that you knit, reduce the number of stitches so that your piece of knitting comes to a point, as follows:

- Row 61: knit every two stitches together to decrease the 40 stitches to 20;
- Row 62: knit every two stitches together decrease the 20 stitches to 10;
- Row 63: knit every two stitches together decrease the 10 stitches to 5;
- Row 64: knit the first two stitches together, then a single stitch, then the next two together to decrease the 5 stitches to 3;
- Row 65: knit all three stitches together into a single final stitch.

Pull out a further 8" (20 cm) of string and cut it off. Pass the end through the last stitch and pull it firmly to tie it off.

03 Put the two sides of your knitting together and sew them together with a needle threaded with the remaining 8" (20 cm) of string. If you catch the last stitch on each row and sew carefully, the join will be almost invisible. When you reach the top, tie a double knot and tuck the end of the string inside.

04 Choose your plants, pot them, and slide the pot inside its 'cozy' (see pp. 54–55).

Retro plants like lots of light and are easy to look after–ideal for 'beginners.' Water them regularly (once a week or every 10 days, depending on the temperature), letting the soil just dry out before each watering.

CACTI IN CUPS

CREATION: ❩❩❩❩❩
CARE: ❩❩❩❩❩
COST: ❩❩❩

Serve up some cacti in a collection of cups for a succulent tea party!

MATERIALS

- 6 assorted teacups and saucers
- A small bag of fine gray gravel
- A small bag of compost for succulents
- Gloves

PLANTS

- 6 small cacti

❩1 Put a little gravel in the bottom of each cup, then some compost, and plant your cacti, adding compost to the required level.

❩2 Add a final layer of gravel and 'tea is served'!

Cacti need very little water and china cups won't allow any surplus to escape, so don't overwater them—and don't water them at all while they're hibernating (see p. 12).

diptyque
MOUSSES

GARDEN GALOSHES

CREATION:)))))
CARE:)))))
COST:)))

MATERIALS

- A pair of rain boots
- 2 heavy-duty plastic bags, preferably the same color as the boots
- Scissors
- Double-sided carpet tape
- A medium-sized bag of indoor plant compost

PLANTS

- 2 different medium-sized plants, e.g. asparagus, ivy, or fern

Rain boots indoors!? If they bring the countryside into your home, why not?

01 Unless you already have a pair you don't use, buy a pair of rain boots. (You can get white ones like those in the picture from companies that supply food manufacturers for around $20–25 / 15–25€.)

02 Fold the plastic bags in half lengthways and slide one into each boot, cutting off the top if it sticks out. The bags should reach down to the sole, but not go into the foot of the boot, and come up to within an inch (3 cm) or so of the top. Peel the protective strip off one side of the double-sided tape and stick it around the inside of each boot, approx. 1" (3 cm) from the top; then peel off the strip on the other side and stick the plastic bag to it.

03 Fill the bags with compost, make a hollow in the top, and plant the specimens of your choice. It's a good idea to have a trailing plant, such as ivy, in one boot and a climbing plant—fern or asparagus—in the other, for contrast. Now, water them.

Water your plants regularly (about once a week), but not too much, as there is nowhere for any surplus water to go. Check the surface of the compost from time to time: it should be just moist.

MESSAGE ON A POT

CREATION: ❩❩❩❩❩
CARE: ❩❩❩❩❩
COST: ❩❩❩

Personalize your plant pots with inspiring or amusing messages—a different one on each pot.

MATERIALS

- 5 or 6 unvarnished plant pots of various sizes
- A roll of masking tape
- A small tin of matte pale pink paint
- A small flat paintbrush
- Letter stamps
- A brown ink pad
- A can of spray varnish
- A bag of indoor plant compost

PLANTS

- An assortment of plants: Aloe vera, succulents, ivy, etc.

❩1 Clean the pots and mark off the areas to be painted with masking tape, if necessary. Alternatively, you can paint the whole pot or paint only thin horizontal lines to accentuate the color of the terracotta.

❩2 Apply one coat of paint and let it dry for 15 minutes or so (paint dries very quickly on terracotta). Then apply a second coat and leave the pot to dry for half an hour.

❩3 Using the letter stamps and the brown ink, print your messages on the painted areas (see fig. 4 on p. 90). Leave to dry for 1 hour.

❩4 Put some compost into the pots and add your plants. Water them generously, as re-potting always 'shocks' the roots.

Water your plants regularly (about once a week), and check the compost from time to time: the surface should be just moist.

I WILL
SURVIVE
TO GROW UP.
GREEN.
MY
PLANT
SUCCULENT

Je cultive Mon balcon
J'♥ les Mauvaises herbes
Je suis Verte.
J'ai la Main verte.

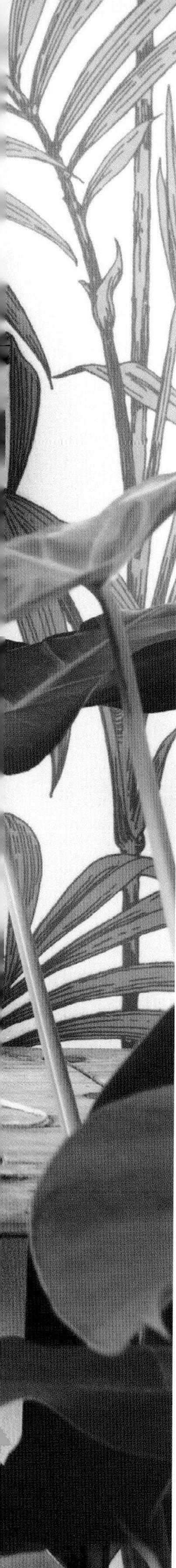

CANNED GREENS

Instead of recycling your cans, why not re-use them as plant pots? After all, greens are good for you!

CREATION: ●●○○○
CARE: ●●○○○
COST: ●○○

MATERIALS

- 4 empty food cans
- Glue remover
- A hammer
- A small sheet of brown paper
- A printer
- A ruler and scissors
- A hole-punch
- White eyelets
- A small bag of indoor plant compost
- A ball of untreated twine

PLANTS

- 4 small plants, e.g. rubber plants or succulents

01 Keep some empty cans and wash them out thoroughly. Tear off the labels and remove any remaining glue with a suitable product.

02 Make a few dents in them by tapping them with a hammer.

03 Print suitable messages on brown paper and cut around them in rectangles. Then make them into labels by cutting off the two corners at one end, punching a hole in between, and sticking an eyelet around it (see fig. 5 on p. 90).

04 Put some compost into the cans, drop in your plants, fill up with more compost, and water well.

05 Cut pieces of string about 2' (60 cm) long and wrap them twice around each can, tying them off with a double knot. Then attach your labels with the loose ends.

If you're using succulents, remember that they don't need much water—and any excess won't be able to escape from the cans. Water them sparingly, every two weeks or so.

ORIGAMI FOR CACTI

CREATION: ❯❯❯❯❯
CARE: ❯❯❯❯❯
COST: ❯❯❯

Wrap your cacti in brown paper and warm words.

MATERIALS

- A roll of brown paper
- A ruler
- Serrated scissors
- Ordinary scissors or a cutter
- A pencil
- (Invisible) glue
- Letter stamps
- A black ink pad
- A fine black ink pen

PLANTS

- 3 small cacti in terracotta pots (c. 3–3.5"/8–9 cm across)

❯1 To make each paper bag, using the instructions below and referring to the figures on p. 91:

- Cut out a 17 × 7" (43 × 18 cm) rectangle of brown paper, the long sides with serrated scissors and the short sides with ordinary scissors or a cutter.
- Divide the rectangle lengthways into three sections by making marks 5.5" (14 cm) and 11" (28 cm) from one end, so that the sections measure 5.5" (14 cm), 5.5" (14 cm), and 6" (15 cm) across (see fig. 6A on p. 91).
- Fold over the longest section (fig. 6B) and make a 1.6" (4 cm) 'tuck' in it (fig. 6C).
- Do the same with the section at the other end and stick the two ends together where they overlap (6D).
- Fold the bag across 1.6" (4 cm) from the bottom, then fold it again 2" (5 cm) higher up (6E).
- Open out the bottom of the bag and fold it in by 1.6" (4 cm) (6F).
- Fold over the top and bottom flaps (6G and 6H) and stick them together where they overlap.
- Add glue as needed to hold the base of the bag firmly together.
- Fold over the top 1.6" (4 cm) of the bag (6I and 6J), taking care not to tear it.

❯2 Fold the bottom of the bag over again so that the bag lies flat, and stamp your message on it (6K), drawing a frame around the message with the ruler.

❯3 Unfold the bag once more and put one of the cacti inside it (see photograph on p. 65).

For the care of cacti, see page 19. When watering, remember to remove the pot from the bag, or any water that runs out of the pot will damage it.

happy:(hap-ee)-n.
feeling of joy
love:(luv)-n.
to be in love

BREAKFAST SPROUTS

Design your own kawaii bowls and serve up a 'breakfast' of bean sprouts!

CREATION:))))))
CARE:))))))
COST:))))

MATERIALS

- 3 small bowls of different colors (e.g. white, cream, and ivory)
- A sheet of tracing carbon paper
- A pencil
- A china marker
- Absorbent cotton
- A spray bottle

PLANTS

- A handful of dried beans

)1 Rinse the beans in warm water and leave them to soak for 24 hours.

)2 Clean the bowls thoroughly to remove any dust or grease.

)3 Trace the 'faces' shown in fig. 7 on p. 92 onto the bowls using the tracing carbon paper and pencil (or draw your own), ink them in with the china marker, and leave to dry for 15 minutes.

)4 Heat your oven to 320 °F (160 °C) and 'bake' your bowls for 25 minutes. Then turn off the oven and leave them inside to cool.

)5 Half-fill the bowls with the cotton, then spray it with water. Put a layer of beans on the cotton: they should start sprouting within a few days.

Spray your beans every day to keep the cotton moist. Once the bean sprouts grow too long and start to droop, remove them, as well as the cotton, and start again.

TREE IN A BAG

CREATION: 2/5
CARE: 3/5
COST: 2/3

Hide your plant pots in a bag, and your interior decor is all sewn up.

MATERIALS

- A piece of thick linen fabric (approx. 3' × 4') (1 × 1.20 m)
- Matching sewing cotton
- Scissors
- A sewing machine
- Number stencils (c. 1"/2.5 cm high)
- Pale pink matte paint
- A paintbrush
- A saucer

PLANTS

- Large retro plant in a pot

01 Make the linen bag in the same way as the paper bags for cacti (see p. 64), referring to the diagrams on p. 91 and following the instructions on p. 92.

02 Paint your lucky number on the bag using number stencils or by copying the numbers in fig. 8 on p. 92, using the carbon tracing paper and a pencil, then painting them in. Leave to dry.

03 Open the bag and put your pot inside it.

If the bag won't stand up by itself, you can stiffen it by spraying starch on it and ironing it in.

1973

Mini GARDENS

Éric Sadin

SUCCULENT PARTERRE

Make yourself a 'crate' little indoor garden with an old wooden box from the market.

CREATION: ❩❩❩❩❩
CARE: ❩❩❩❩❩
COST: ❩❩❩

MATERIALS

- 1 small fruit/ vegetable crate
- A roll of clear plastic film
- A staple gun
- A large bag of cacti and succulent compost
- A few handfuls of fine gray gravel (aquarium gravel is ideal)

PLANTS

- 6–10 succulents (depending on size) of various kinds, e.g. species of houseleek (*Sempervivum*)

01 Cut a rectangle of plastic film a little larger than your crate and lay it inside. Staple it firmly to the crate—first to the bottom and then to the sides. Fold the plastic in the corners to keep it flat. Trim off any excess at the top.

02 Fill the crate with a layer of compost just over 1" (3 cm) deep.

03 Make small hollows for your succulents and plant them.

04 Finally, sprinkle gravel between the plants and water them in.

Succulents like dry conditions, so water them lightly, every two weeks or so. The compost should never be wet, or your plants might rot.

HANGING HERB GARDEN

CREATION: ❩❩❩❩❩
CARE: ❩❩❩❩❩
COST: ❩❩❩

Always have herbs at your fingertips with this handy 'wall garden.'

MATERIALS

- A piece of coarse natural linen fabric (5' × 4')/ 1.5 m × 1.20 m
- Matching sewing cotton
- Scissors
- A sewing machine
- An iron
- A pencil
- Pins
- 1 skein of black embroidery thread
- An embroidery needle
- 6 freezer bags
- A few handfuls of indoor plant compost

PLANTS

- 6 small herb plants, e.g. (flat) parsley, basil, coriander, or thyme

1 Make the herb holder by referring to figs. 9A–D and following the instructions on p. 93.

2 Embroider a number on each pocket:

- Measure 1.4" (3.5 cm) from the side of the pocket and 2" (5 cm) down from the top and make a pencil mark. Untwist the skein of embroidery thread and cut a single length of about 2.6' (80 cm). Embroider in cross point, i.e. crossing two strands at a time.
- To make the hanging loops, cut a 2.6' (80 cm) length of all eight threads in the skein together. Knot one group of ends together and thread the other through the needle. Push the needle through the linen about 1" (2.5 cm) from one of the top corners, then back through it ½" (1.25 cm) further in, leaving a loop of thread. Double the thickness of the loop by pushing the needle back through the linen where you started, make a knot, and cut off the spare thread. Do the same at the opposite corner.

3 Now it's time to put the herbs in the pockets. Fold each freezer bag in half lengthways, ensuring that it will fit snugly into a pocket, and put a little compost in the bottom of it. Plant one of the herbs in the bag and slide it into its pocket. Repeat the process for the other five pockets.

4 Your herbs should be hung where there's plenty of light, but not against a window.

Once or twice a week (depending on the temperature), pour a little water gently into each bag.

01
02
03
04
05
06

MAUVAISE
GRAINE

AN OLD CRATE 'HEAVEN-SCENT'

Plant a recycled crate with aromatic herbs to make a natural air-freshener.

CREATION:))

CARE:))

COST:))

MATERIALS

- An old apple crate
- Letter stencils
- A small tin of pale pink paint
- A paintbrush
- 1 sheet of fine sandpaper
- A roll of clear plastic film
- A staple gun
- A large bag of indoor plant compost

PLANTS

- An assortment of herbs, e.g. (flat) parsley, basil, coriander, and thyme

1 Find an old wooden apple crate and brush off any dirt and dust.

2 Mark one side with your wording of your choice, using letter stencils about 2" (5 cm) high, working the paint into the wood with your brush. When the paint is dry, rub it gently with fine sandpaper so that the wording looks as old and worn as the rest of the crate.

3 Measure the crate and cut a rectangle of plastic film to fit, as follows:

- Length = length of crate + twice the height of the crate - 3" (8 cm) (because the edge of the film should be 1½" (4 cm) below the top of the crate so that it isn't seen);
- Width = width of crate + twice height of crate - 3" (8 cm).
- Lay the plastic in the crate and staple it first to the bottom, then to the sides, folding it at the corners to keep it flat.

4 Fill the crate to within 1½" (4 cm) of the top, make small hollows for your herbs, and plant them, watering them well when you've finished.

Water your herbs slightly about once a week, remembering that if you give them too much water, the excess won't be able to drain away.

GARDEN IN A JAR

CREATION:)))))
CARE:)))))
COST:)))

Turn a jam jar into a tropical paradise faster than you can say 'epiphyte.'

MATERIALS

- 1 large clip-top glass jar
- A few handfuls of fine gray gravel
- A small bag of water-retaining compost (generally containing peat)

PLANTS

- 1 small fern
- 1 small epiphyte

)1 Clean your jar thoroughly—to remove any residues if it's old, or any chemicals if it's new.

)2 Pour gravel into the bottom to make a 0.8" (2 cm) layer.

)3 Add a 0.8" (2 cm) layer of compost and make a hollow in it just off-center.

)4 Remove as much soil as possible from the roots of the fern, taking care not to damage them, and plant it inside the jar, pressing the compost down around it.

)5 Add a second 0.8" (2 cm) layer of gravel, which will help with drainage, in addition to stabilizing the fern in its shallow bed of compost.

)6 Place the epiphyte on top of the gravel beside the fern, giving both plants a little water.

To prevent a build-up of bacteria, air must be allowed to circulate inside the jar, so leave the lid slightly open. Place the jar in a shady spot and occasionally give it just a little water.

(MENTHE)
ROMARIN*
(THYM)
*PERSIL

HERBAL 'PALETTE'

Make an impression with this herb garden on your window ledge—for very little 'monet'!

CREATION: ❩❩❩❩❩ (2 of 5)
CARE: ❩❩❩❩❩ (3 of 5)
COST: ❩❩❩ (1 of 3)

MATERIALS

- 1 wooden pallet
- A saw
- A claw
- A hammer
- A screwdriver and 4 screws
- A large can of off-white matte paint
- A sheet of letter transfers
- A roll of clear plastic film
- A staple gun
- A large bag of potting compost

PLANTS

- An assortment of herbs, e.g. thyme, coriander, rosemary, parsley, and basil

1 Saw the pallet according to the instructions on p. 95 (figs. 11A & B).

2 Give it two coats of paint, leaving it to dry for 2 hours between coats and following the instructions on the container.

3 Apply the names of your selected herbs 0.6" (1.5 cm) from the top of the wood, using the letter transfers, as shown in the photograph on p. 80.

4 To retain the moisture, cut two rectangles of clear plastic film, one for each pallet section, as follows:

- Length = length of base + twice the height of the sides;
- Width = width of base + twice the height of the sides.

Lay the film inside the pallet and staple it to the wood, first around the bottom, then up the sides, folding the film in the corners to keep it tidy.

5 Fill each section with compost, make hollows for your herbs, and plant them. Pack the soil in and water well.

Water your herbs once or twice a week, according to the weather and the temperature. Hardy herbs (e.g. rosemary and mint) should survive the winter, but annuals (coriander, aniseed, etc.) will need to be replaced every spring.

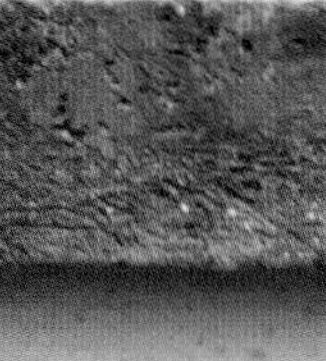

SEED SHELLS

CREATION: ●○○○○
CARE: ●○○○○
COST: ●○○

Delight your senses with this tiny fairyland of 'hatching' seeds.

MATERIALS

- 4–6 white eggs
- Absorbent cotton
- A spray bottle

PLANTS

- A few dried seeds: beans, peas, lentils, etc.

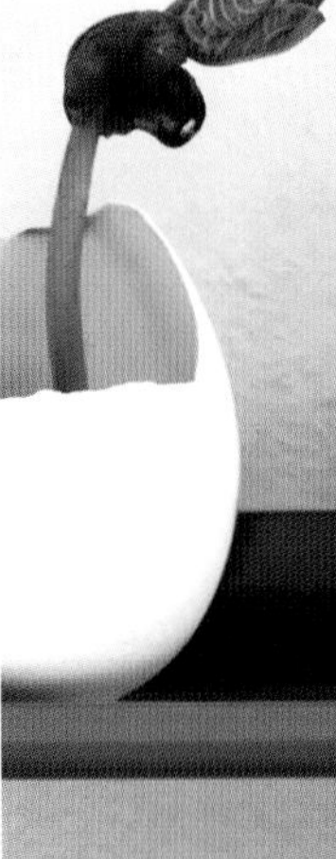

1 Rinse the seeds in warm water and leave them to soak for 24 hours.

2 Break the eggs sharply, about one-third of the way down, and remove the contents (which, of course, you can use to make a delicious omelet!). Wash the shells thoroughly and leave them to dry.

3 Lay some cotton at the bottom of each egg shell, filling about one-third of it.

4 Spray water on the cotton and place one seed in each shell. They should germinate within a few days.

Spray a little water on the seeds every day to keep the cotton moist.

GREEN GLASSES

Return to your childhood and indulge in the pleasure of growing something from (almost) nothing.

CREATION: ❩❩❩❩❩
CARE: ❩❩❩❩❩
COST: ❩❩❩

MATERIALS

- An assortment of old glasses
- Toothpicks
- Small plant pots
- A small bag of interior plant compost
- Secateurs
- Gloves

PLANTS

- An avocado, a potato, a pineapple, some ivy, and a bramble

❩1 **Avocado:** Remove the seed from an avocado without damaging it with the knife, remembering which part of it was at the pointed end of the avocado, keeping this at the top. Stick three toothpicks a little way into it from the sides about halfway up. Set the seed, still 'upright,' on a glass and fill it with water until the bottom third of the seed is covered. Put it by a window.
After about a month, a taproot will appear under the water and the top of the stone will split, allowing a shoot to emerge.

❩2 **Potato:** Use an old potato (preferably an organic one) that has started to sprout, and simply plant it in a pot of compost. Water it regularly and you'll soon see a shoot and leaves appear. Once the stem is about 6" (15 cm) tall, you can pot the potato.

❩3 **Pineapple:** Choose one with a good crown. Remove this by twisting it with your hands, leaving the fruit itself intact. Carefully remove the lowest leaves and gently stick four toothpicks into the base. Set it on a glass with a little water in the bottom, and place it by a window.
After a few weeks, tiny white roots will appear. When these have grown properly, fill a pot with compost and plant your pineapple 'tree.' Water it about once a week. It will grow—but very slowly!

❩4 **Ivy:** Cut a stem of ivy with a pair of secateurs and place it in a glass of water. Roots should appear within a few weeks.

❩5 **Bramble:** When you're out walking, gently pull up a small bramble (be sure to wear gloves and mind the thorns!), so that the roots come with it. Put it in a glass of water and it will be quite happy. As it grows, you can either pot it or leave it where it is!

Templates and instructions

IVY TRELLIS

(p. 30)

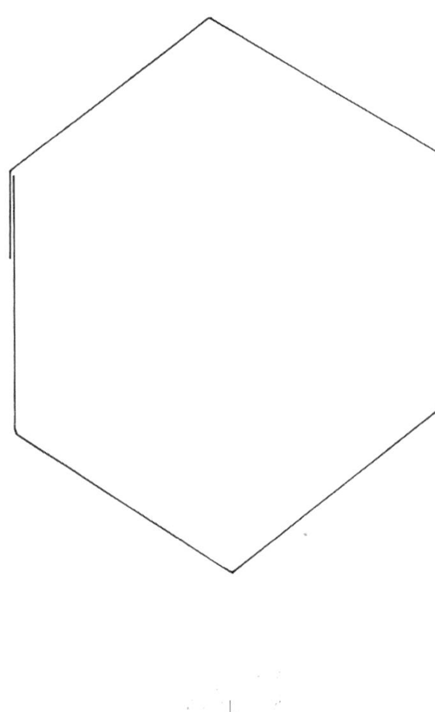

fig.1

GREEN DREAM-CATCHER

(p. 35)

fig. 2A

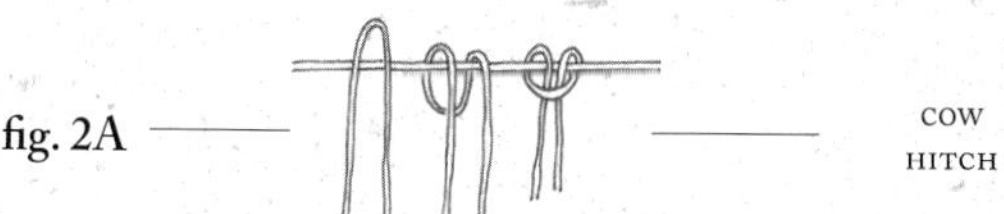

COW HITCH

- Fold the thread in half and pass it around the branch as shown above.

fig. 2B

SQUARE KNOTS

STRAIGHT THREAD

GRANNY KNOTS

- Take one thread from each 'pair' and tie them together using a square knot to form triangles with ½" (1.5 cm) sides, as shown above and in the main diagram.
- Use the central 6 threads to make the pot-holder. Then make the sides so that they match. The straight threads should be 2" (5 cm) long. Then tie 4 threads together using granny knots.
- Continue as shown in the diagram and cut the threads off in a straight line at the bottom.

For more information on how to tie the knots, see next page.

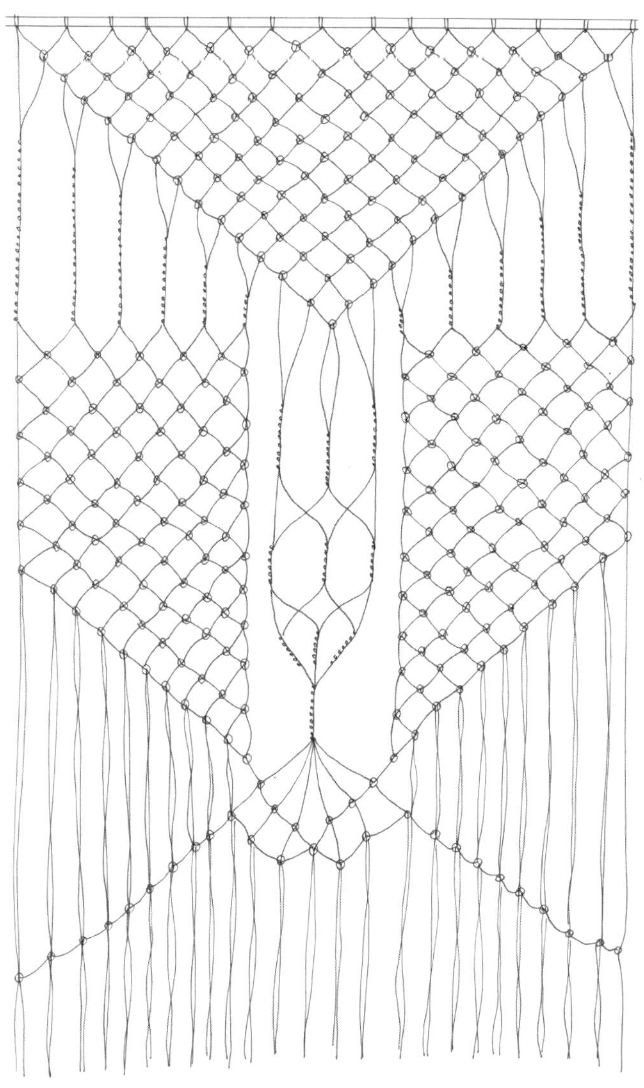

NATURAL MACRAMÉ

(p. 41)

How to tie the knots

- **Square knots** (fig. 3A) are used to make a flat weave. The two halves of the knot are made in *opposite* directions, e.g. first passing the left thread over the right and then the right over the left.
- **Granny knots** (fig. 3B) are used to make a turning or twisting weave. The two halves of the knot are made in the *same* direction. You can make a twisting weave with any even number of threads: e.g. with 4 threads, tie the two central threads to the two outer threads; with 6 threads, tie the two central threads to the four outer threads.
- **Heart shapes** (fig. 3C) add interest to otherwise straight threads between two knots. The ropes are free and mingle.

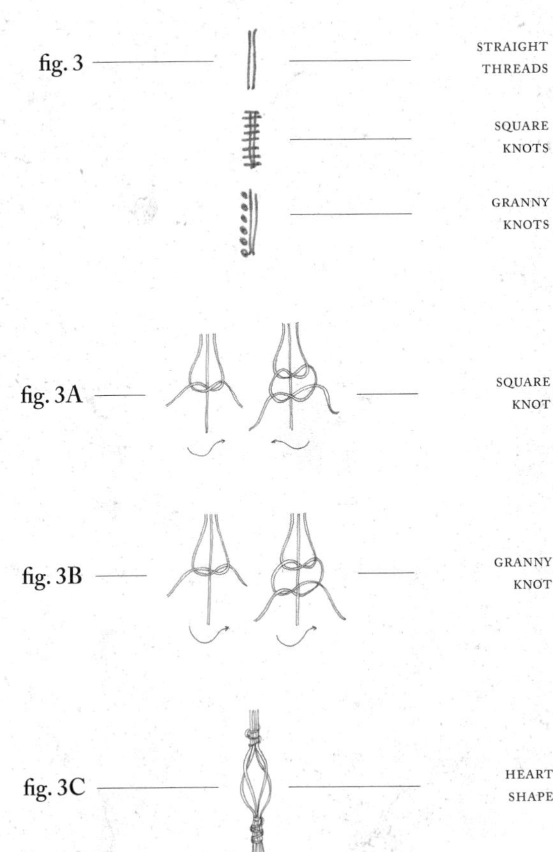

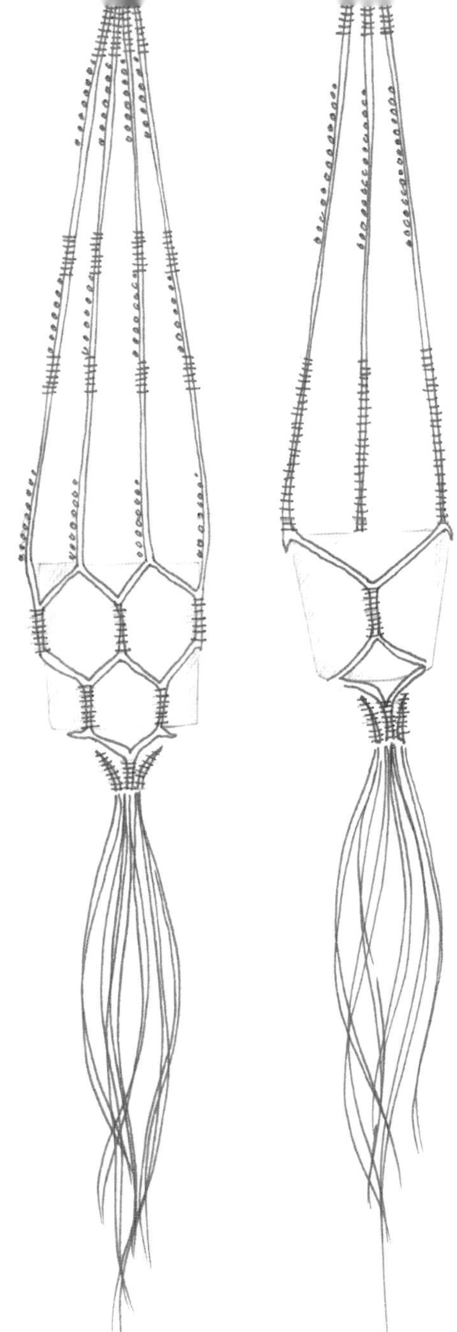

- Cut 11 threads 10' (3 m) long and 1 thread 11'6" (3.5 m) long. Cut the remainder of the long thread so that it is the same length as the others. Wind the long thread tightly around the others to make a 3" (8 cm) 'binding'. Bend this in half and wind the long thread 5 times around all 24 threads, tying a knot each time, to make a loop for hanging up your basket.
- Decide whether to make a 3-cord or a 4-cord hanging basket, as shown above.

4-cord version

- Divide the 24 threads into four groups of 6.
- In each group of threads, tie knots as described below to make the pattern shown in the diagram opposite. For each knot, tie the 2 central threads to the 2 outer threads.
  - 2.4" (6 cm) of square knots;
  - 1.2" (3 cm) of straight threads;
  - 3.5" (9 cm) of granny knots;
  - 2.4" (6 cm) of straight threads;
  - 3" (8 cm) of square knots;
  - gather the threads: divide the cords in two to join them with the middle of the adjacent cord;
  - 2.4" (6 cm) of straight threads;
  - 0.8" (2 cm) of square knots;
  - divide the cords in two again;
  - 2" (5 cm) of straight threads;
  - 0.8" (2 cm) of square knots;
  - divide the cords in two again;
  1.6" (4 cm) of straight threads;
  0.4" (1 cm) of square knots;
  - divide the cords in two again;
  0.4" (1 cm) of straight threads;
  0.8" (2 cm) of square knots;
  - tie the 4 groups of threads together, preserving their length, do not trim them with scissors.

3-cord version

- Divide the 24 threads into three groups of 8.
- In each group of threads, tie knots as described below to make the pattern shown in the opposite diagram. For each knot, tie the 4 central threads to the 4 outer threads.
  - 2.4" (6 cm) of square knots;
  - 2.8" (7 cm) of granny knots;
  - 3.5" (9 cm) of straight threads;
  - 2" (5 cm) of square knots;
  - 3" (8 cm) of granny knots;
  - 0.8" (2 cm) of square knots;
  - 1.6" (4 cm) of straight threads;
  - 2" (5 cm) of granny knots;
  - gather the threads: divide the cords in two to join them with the middle of the adjacent cord;
  - 3" (8 cm) of straight threads;
  - 0.8" (2 cm) of square knots;
  - divide the cords in two again;
  - 1.2" (3 cm) of straight threads;
  - 0.4" (1 cm) of square knots;
  - tie the 3 groups of threads together, preserving their length, do not trim them with scissors.

MESSAGE ON A POT *(p. 60)*

fig. 4

Some messages you might like to use

I WIll
SURVIVE

GROW STRONG.

URBAN
JUNGlE

GREEN.

MY
SUCCULENT
PlANT

CANNED GREENS *(p. 63)*

Label template

fig. 5

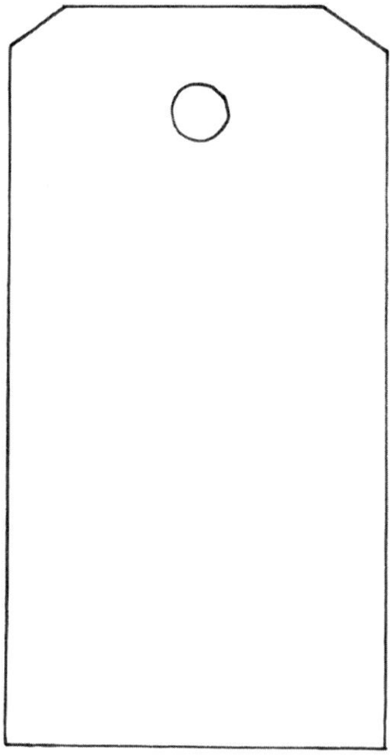

ORIGAMI FOR CACTI

(p. 64)

fig. 6

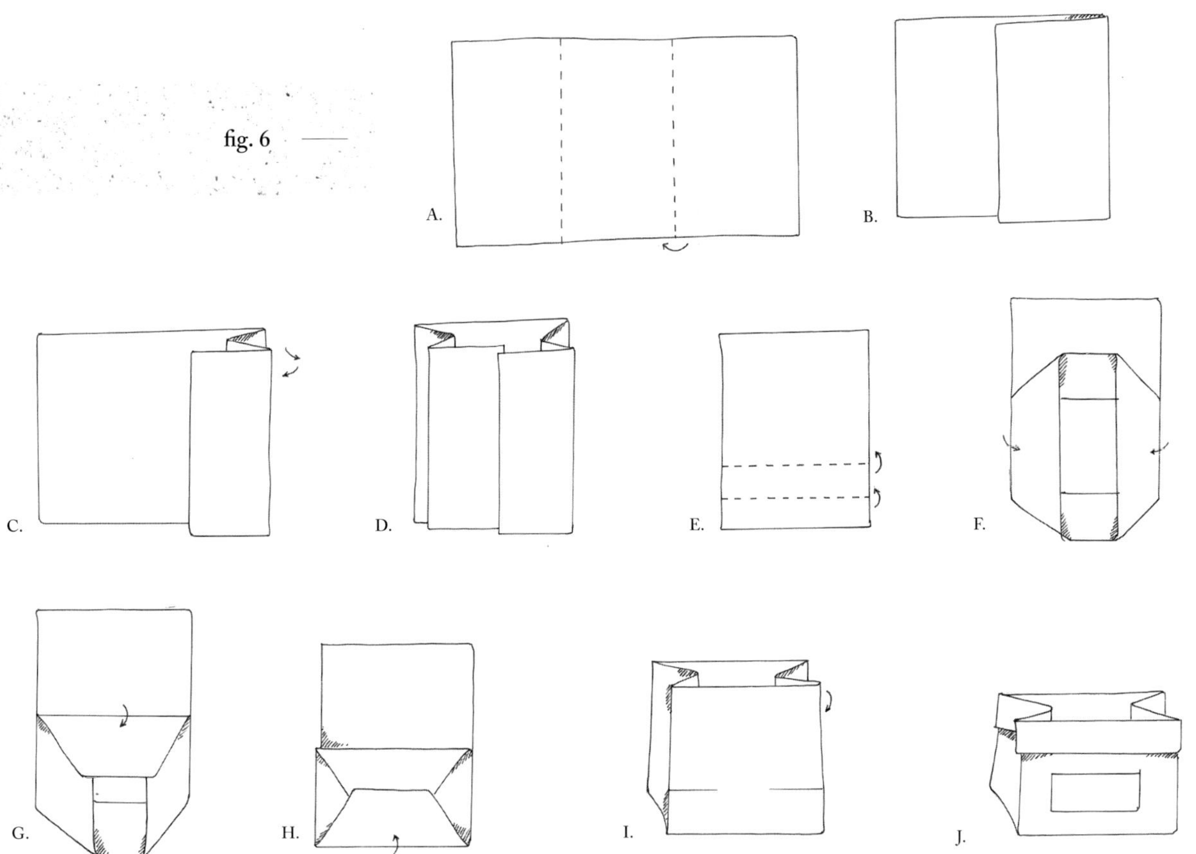

life: (lahyf) -n.
to be full of life

fig. 6K

Some messages you might like to use

love: (luv) -n.
to be in love

fig. 7

BREAKFAST SPROUTS
(p. 67)

TREE IN A BAG
(p. 68)

How to make the bag

- Cut a rectangle of linen 46 × 30" (115 × 76 cm). One of the long sides should be the edge of the piece of fabric; make this the top of your bag to give it a suitably 'rough' look (don't make a nice neat hem!).
- Measure and mark two folds to make three 'panels' of 15" (38 cm), 15" (38 cm), and 16" (41 cm) (as shown in fig. 6A)
- Fold over the longest panel (fig. 6B) and make a 4" (10 cm) 'tuck' in it (6C).
- Do the same with the opposite panel (6D) and sew the overlapping panels together: put the 1" (2.5 cm) overlap on each side together, right side to right side, and sew along the join.
- Mark the fabric 8" (20 cm) and 16" (40 cm) from the bottom (6E).
- Open up the bag and fold the two sides into the center (6F).
- Fold the two flaps in (6G & 6H).
- Sew up the bottom of the bag: one strong stitch at each end of the overlap should be enough.

Number template

fig. 8

HANGING HERB GARDEN

(p. 74)

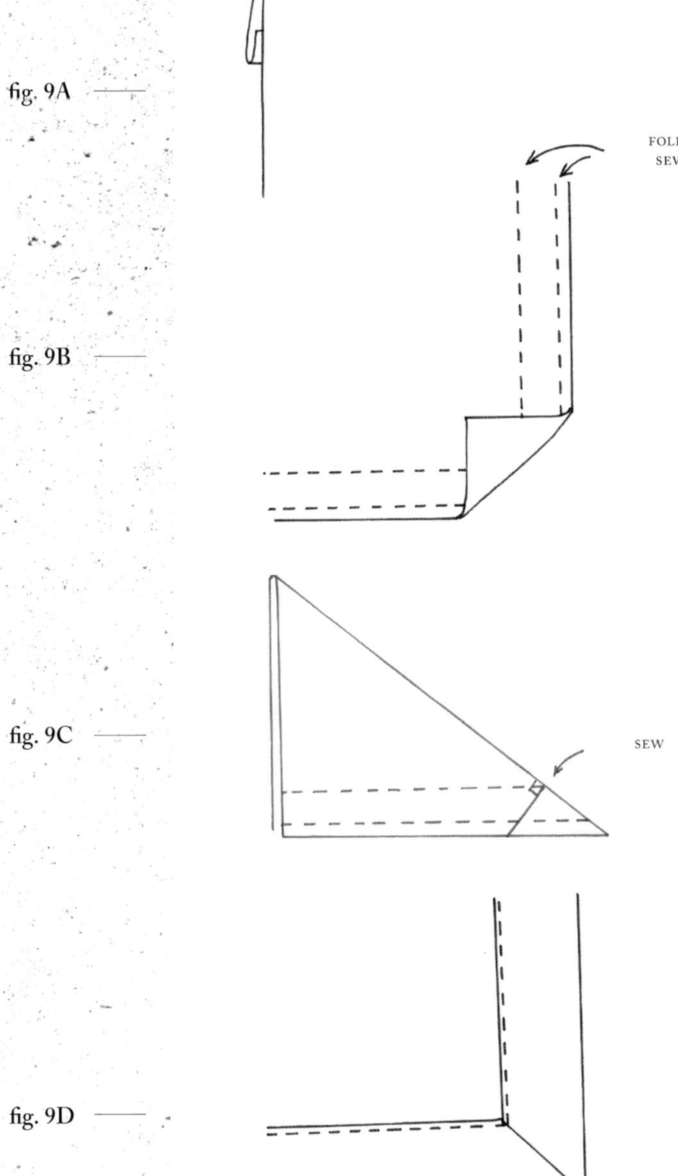

How to make the wall-hanger

- Cut out a rectangle of fabric measuring 27 × 18" (68 × 41 cm) (measurements include hems).
- Hem it all the way around, ironing creases ½" (1.3 cm) and 1½" (4 cm) from the edges, making a double fold along each side, and folding in the corners, as shown in figs. 9A–D. The rectangle should now measure 23 × 14" (58 × 36 cm).
- Then cut out two rectangles measuring 13 × 10" (26 × 35 cm). Hem one of the long sides of each rectangle in the same way as the large rectangle (see above). Make a single ½" (1.3 cm) fold along the other three edges and hem these, too. You should now have two rectangles measuring approximately 12 × 7½ " (21 × 33 cm).
- Lay one of the smaller rectangles across the top half of the large rectangle, leaving a 4" (10 cm) gap at the top and 1" (2.5 cm) gaps at each side. Pin it to the large rectangle. Then position the other small rectangle below it, leaving a 1" (2.5 cm) gap at the sides and the bottom; there should be a 3" (8 cm) gap between the two small rectangles. Sew the two small rectangles in position by stitching around three sides, leaving the tops open.
- Divide each small rectangle lengthwise into three equal sections (4" (10 cm) across), marking the divisions with a pencil and then sewing along them. This will give you six pouches for your herbs (see p. 75).

AN OLD CRATE 'HEAVEN-SCENT' *(p. 77)*

fig. 10

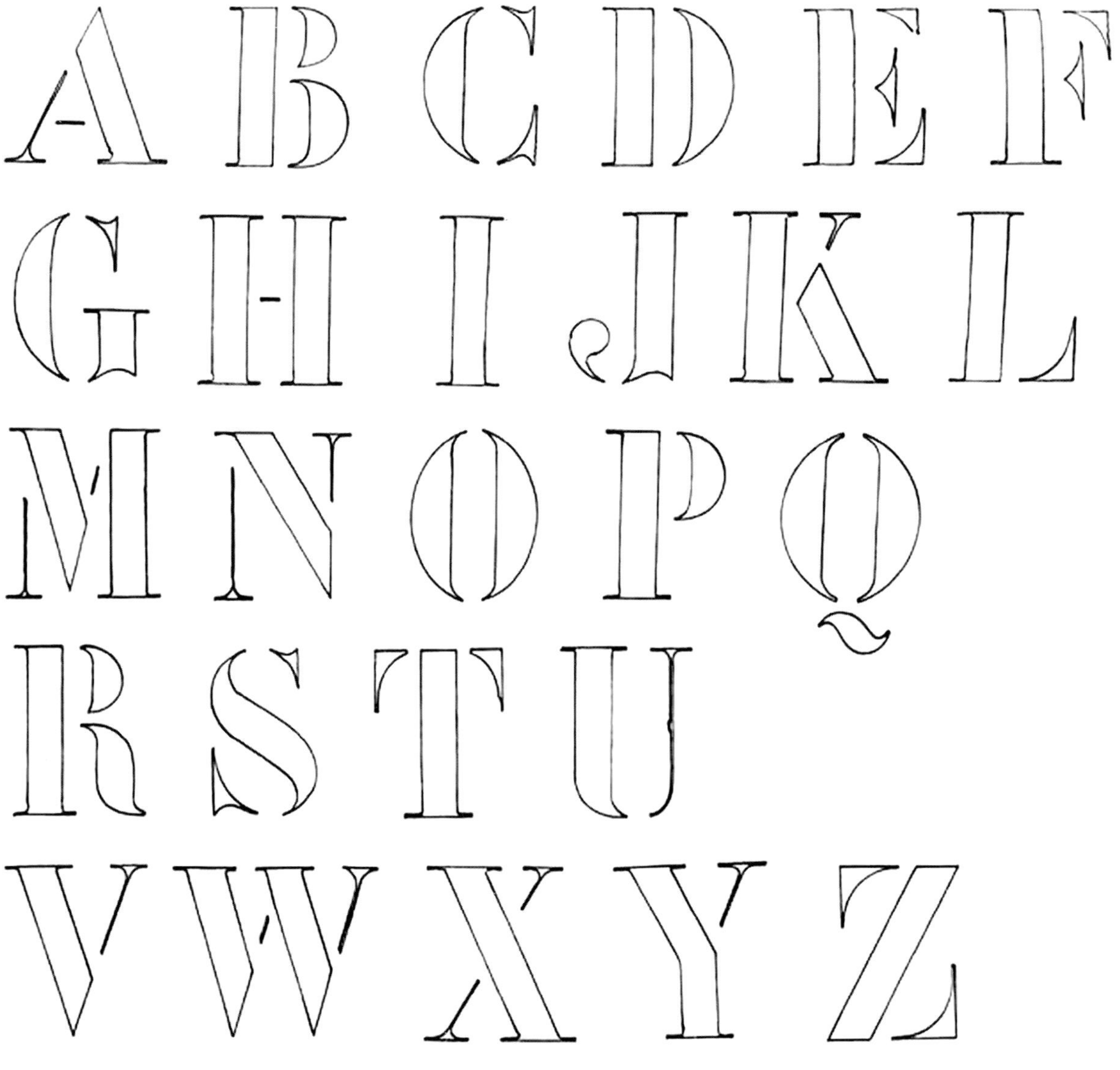

A B C D E F
G H I J K L
M N O P Q
R S T U
V W X Y Z

HERBAL 'PALETTE'

(p. 81)

fig. 11A

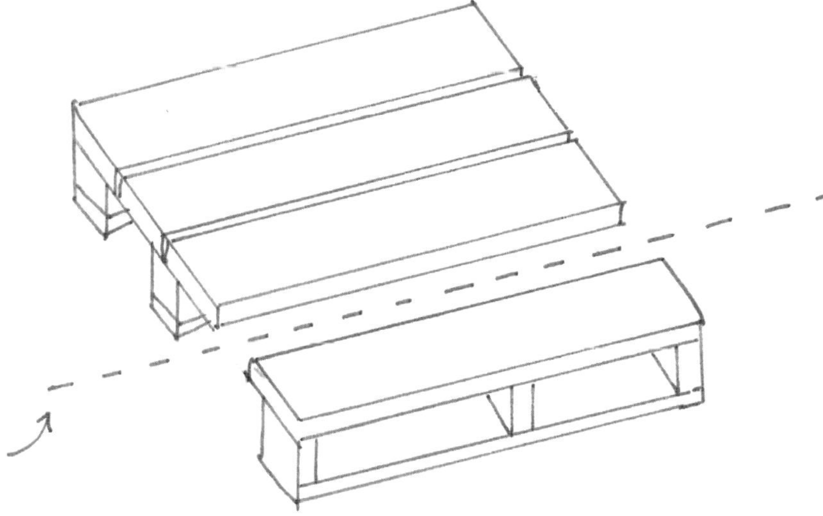

- Saw off one section of the pallet as shown in fig. 11A to make your herb trough.
- Remove a plank from the rest of the pallet to form the base of your trough. Pry it off with the claw; then knock out the nails with a hammer.

- Screw the plank to the underside of your trough (fig. 11B).

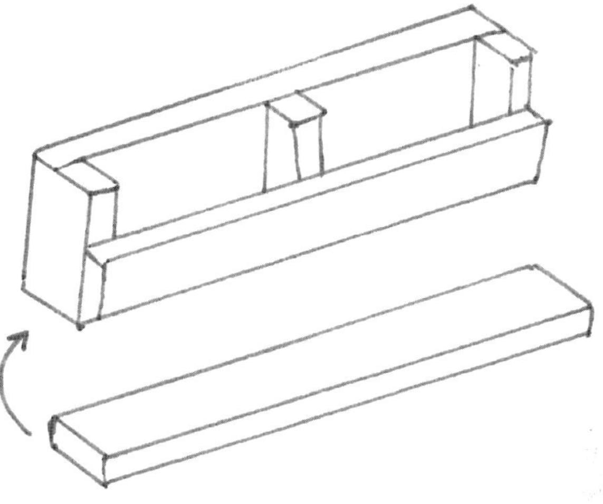

fig. 11B

Imprint

Published in 2019 by:

GINGKO PRESS

Gingko Press Verlags GmbH
Schulterblatt 58
D-20357 Hamburg
Germany
Tel: +49 (0)40-291425
Fax: +49 (0)40-291055
Email: gingkopress@t-online.de

Gingko Press Inc
2332 Fourth Street, Suite E
Berkeley, CA 94710
USA
Tel: (510) 898 1195
Fax: (510) 898 1196
Email: books@gingkopress.com
www.gingkopress.com

ISBN: 978-3-943330-26-7

Translation From the French: Joseph Laredo © Gingko Press Verlags GmbH
Copyediting: John Stilwell
Typesetting: Weiß-Freiburg GmbH — Graphik und Buchgestaltung
Project Coordination: Anika Heusermann
Printed in Spain by Grafica Estella

Published originally under the title:
Home Jungle – Invitez les plantes à la maison

For the original French edition:
Director: Jean-Louis Hocq
Editorial Director: Corinne Cesano
Editors: Anne Kalicky, Comptoir Éditorial
Editorial Collaboration: Mathilde Poncet
Copy-Editing: Clémentine Sanchez
Design: Le Bureau des Affaires Graphiques
Frabrication: Laurence Duboscq
Reproduction: APS

Also available from
Gingko Press

ISBN: 978-1-58423-713-6